# TRANSITIONAL M...
## Developing Number Sense

# ASSESSMENTS BOOK

John Woodward, Ph.D. and Mary Stroh, B.S.

Sopris West Educational Services • Longmont, Colorado

Copyright 2004 by Sopris West Educational Services
All rights reserved.

Permission is granted to the purchaser to photocopy the student quizzes and tests for use in his or her classroom only. No other part of this work may be reproduced or transmitted in any form or by any means, electronic or mechanical, including photocopying or recording, or by any information retrieval system, without the express written permission of the publisher.

ISBN 1-57035-963-6

*Edited by* Beverly Rokes
*Editorial assistance by* Annette Reaves
*Text design by* Edward Horcharik *and* Sebastian Pallini
*Text production by* Edward Horcharik *and* Matthew Williams
*Cover design by* Sue Campbell
*Production assistance by* Denise Geddis

06 05 04 03     6 5 4 3 2 1

Printed in the United States of America

Published and Distributed by

SOPRIS
WEST
EDUCATIONAL SERVICES

4093 Specialty Place • Longmont, Colorado 80504 • (800) 547-6747
www.sopriswest.com

212ASSESS/11-03/BAN/4M/070

# ASSESSMENTS
## Quiz

Name _____  Date _____

## PART 1
*Give the value of the digit that is underlined and in bold type.*

1. **3**24 _____

2. **5**,096 _____

3. 5**1**2 _____

4. 72**0** _____

## PART 2
*Write the following numbers in expanded form.*

1. 217 _____

2. 375 _____

3. 602 _____

4. 450 _____

## PART 3
*Solve the following basic and extended facts.*

1. 7 + 8 = ____     70 + 80 = ____     700 + 800 = ____

2. 9 + ____ = 15     ____ + 90 = 150     1,500 = ____ + 900

3. 12 = 7 + ____     70 + ____ = 120     700 + ____ = 1,200

Assessments • Number Sense                          Unit 1 Quiz    1

# PART 4

*Write the following in expanded form and solve. Write the answer the regular way.*

1.  225 →   + ___|___|___   Answer: ____
    + 314

2.  637 →   + ___|___|___   Answer: ____
    + 151

# ASSESSMENTS
Test

Name _____  Date _____

## PART 1
*Give the value of the digit that is underlined and in bold type.*

1. 250,0**1**9 _____
2. 3,24**7** _____
3. 2,**4**39 _____
4. 1,02**8**,019 _____

## PART 2
*Write the following numbers in expanded form.*

1. 347 _____
2. 429 _____
3. 508 _____
4. 917 _____

## PART 3
*Solve the following basic and extended facts.*

1. 9 + 2 = ____      90 + 20 = ____      900 + 200 = ____
2. 8 + ____ = 15      80 + ____ = 150      1,500 = 800 + ____
3. 14 = ____ + 6      60 + ____ = 140      ____ + 600 = 1,400
4. 130 = 60 + ____      6 + ____ = 13      1,300 = 600 + ____
5. 120 = 50 + ____      500 + ____ = 1,200      5 + ____ = 12

# PART 4

*Write the following problems in expanded form and solve. Then write the answer the regular way.*

1. 523
   + 143       +            Answer: ____

2. 721
   + 207       +            Answer: ____

3. 348
   + 417       +            Answer: ____

4. 409
   + 381       +            Answer: ____

# PART 5

*In the following word problem, underline the important information and tell what the problem is asking. Then solve. Use your calculator if needed.*

**Problem:** There were 325 students at the Homecoming dance and 275 students at the Winter Festival dance. The president of the Student Council wanted to know the total number of students at the 2 events. There is also a dance coming up in the spring.

What is the problem asking?

Answer _____

# PART 6
*Some of these problems have errors. Find the errors and then correct them.*

1.  267  
   + 24  
   ‾‾‾  
    281

2.  316  
   + 49  
   ‾‾‾  
    355

3.  275  
   + 394  
   ‾‾‾‾  
    669

4.  478  
   + 229  
   ‾‾‾‾  
    607

# ASSESSMENTS
## Quiz

Name _____  Date _____

## PART 1
*Write the fact family for the following numbers.*

**1.** 9, 2, and 11      **2.** 7, 6, and 13      **3.** 5, ?, and 14

## PART 2
*Write the following in expanded form and solve.*

**1.**  68  →         Answer ____
       − 34

**2.**  197 →         Answer ____
       −  62

**3.**  579 →         Answer ____
       − 126

## PART 3
*Indicate the best rounding strategy for each problem. Then estimate the problems using the strategies selected. Finally, find the exact answer on your calculator and compare it to your estimate.*

| Problem | Rounding Strategy | Estimate | Calculator Answer |
|---|---|---|---|
| 1.  801<br>− 209 | | | |
| 2.  76<br>− 57 | | | |
| 3.  789<br>− 111 | | | |
| 4.  874<br>− 526 | | | |

## PART 4
*Solve using any method.*

1.  91
    − 54

2.  782
    − 91

3.  821
    − 419

4.  423
    − 314

# ASSESSMENTS
Test

Name _____ Date _____

## PART 1
*Solve the following basic and extended subtraction facts.*

1. $17 - 9 =$ ____    $170 - 90 =$ ____    $1{,}700 - 900 =$ ____
2. $16 -$ ____ $= 8$    $160 -$ ____ $= 80$    $1{,}600 - 800 =$ ____
3. ____ $- 9 = 5$    ____ $- 90 = 50$    ____ $- 900 = 500$
4. $13 - 7 =$ ____    $130 -$ ____ $= 70$    $1{,}300 - 700 =$ ____

## PART 2
*Write the addition/subtraction fact families for the following basic and extended facts.*

1. 2, 9, and 11

3. 7, 8, and 15

2. 20, 90, and ?

4. 70, ?, and 150

# PART 3
*Write the following in expanded form and solve.*

1.  97  →                    Answer ____
   − 45

2.  879 →                    Answer ____
   − 58

3.  358 →                    Answer ____
   − 127

# PART 4
*For each problem, state the best rounding strategy to use, estimate the answer using that strategy, and then find the exact answer on your calculator and compare it to your estimate.*

| Problem | Rounding Strategy | Estimate | Calculator Answer |
|---|---|---|---|
| 1. 73 − 26 | | | |
| 2. 498 − 299 | | | |
| 3. 324 − 199 | | | |
| 4. 901 − 602 | | | |

## PART 5
*Solve the following using any method.*

1.  62
    − 39

2.  991
    − 87

3.  741
    − 381

4.  561
    − 157

## PART 6
*Some of the following problems contain errors. Find the errors and fix them.*

1.  46
    − 27
    ----
    21

    Does this problem have an error? _____

    If so, write the correct answer here: _____

2.  562
    − 43
    -----
    519

    Does this problem have an error? _____

    If so, write the correct answer here: _____

3.  619
    − 427
    -----
    212

    Does this problem have an error? _____

    If so, write the correct answer here: _____

4.  785
    − 376
    -----
    409

    Does this problem have an error? _____

    If so, write the correct answer here: _____

# PART 7
*Choose the best method to solve each of the following problems, then solve the problems using that method. Write the method you chose. The choices are:*

(a) Traditional Subtraction

(b) Estimation and Calculator

(c) Good Number Sense and Number Line

1. 444  Method:
   − 222


2. 400  Method:
   − 398


3. 1,012  Method:
   − 574

# ASSESSMENTS
## Quiz

Name _____  Date _____

## PART 1
*Solve the following basic and extended multiplication facts.*

1. $6 \times 8 =$ ____      $6 \times 80 =$ ____      $6 \times 800 =$ ____
2. $9 \times 7 =$ ____      $9 \times 70 =$ ____      $9 \times 700 =$ ____
3. $8 \times 7 =$ ____      $8 \times 70 =$ ____      $8 \times 700 =$ ____
4. $5 \times 9 =$ ____      $5 \times 90 =$ ____      $5 \times 900 =$ ____

## PART 2
*Pull out the 10 from each of the following numbers.*

**EXAMPLE**     $500 = 50 \times 10$

1. $40 =$ ____      4. $400 =$ ____      7. $4{,}000 =$ ____
2. $60 =$ ____      5. $600 =$ ____      8. $6{,}000 =$ ____
3. $500 =$ ____     6. $50 =$ ____       9. $5{,}000 =$ ____

# PART 3
*Fill in the missing expressions in the table.*

| The Starting Number | ? × 1,000 | ? × 100 | ? × 10 |
|---|---|---|---|
| 5,000 | 5 × 1,000 |  | 500 × 10 |
| 7,000 |  | 70 × 100 |  |
| 9,000 |  |  | 900 × 10 |
| 6,000 |  | 60 × 100 |  |

# PART 4
*Write the following multiplication problems in expanded form. Solve.*

1.  39
    × 7

2.  78
    × 6

3.  65
    × 9

4.  39
    × 8

# ASSESSMENTS
## Test

Name _____  Date _____

## PART 1
*Solve the following basic and extended multiplication facts.*

**1.** 7 × 6 = ____      7 × 60 = ____      7 × 600 = ____

**2.** 9 × 8 = ____      9 × 80 = ____      9 × 800 = ____

**3.** 8 × 7 = ____      8 × 70 = ____      8 × 700 = ____

**4.** 6 × 9 = ____      6 × 90 = ____      6 × 900 = ____

## PART 2
*Pull out the 10 from each of the following numbers.*

**EXAMPLE**     50 = 5 × **10**

**1.** 60 = ____      **3.** 80 = ____      **5.** 150 = ____

**2.** 20 = ____      **4.** 90 = ____      **6.** 300 = ____

## PART 3
**Complete the following problems about the metric system.**
*Review:*

10 millimeters (mm) = 1 centimeter

10 centimeters (cm) = 1 decimeter

10 decimeters (dm) = 1 meter

1. 2 mm × 10 = _____ mm       or       _____ cm
2. 5 cm × 10 = _____ cm       or       _____ dm
3. 10 dm × 10 = _____ dm     or       _____ m
4. 6 mm × 10 = _____ mm       or       _____ cm
5. 8 cm × 10 = _____ cm       or       _____ dm

## PART 4
*Write the following in expanded form and solve.*

1. 97 →
   × 6      + _____ | _____

2. 85 →
   × 9      + _____ | _____

## PART 5
*Fill in the missing expressions in the table.*

| The Starting Number | ? × 1,000 | ? × 100 | ? × 10 |
|---|---|---|---|
| 1,000 | | | |
| 2,000 | | | |
| 3,000 | | | |
| 6,000 | | | |
| 7,000 | | | |

## PART 6
*Solve the following problems using the traditional method of multiplication. Check your answers with estimation and a calculator.*

1.  96
    × 3

2.  84
    × 4

3.  72
    × 9

## PART 7

*Solve the following problems by first using estimation to get a good ballpark figure, and then using your calculator to get an exact answer.*

> EXAMPLE
> 49 × 53
> Estimate: 50 × 50 = 2,500
> Exact Answer: 2,597

1. 67 × 8        Estimate _____        Exact Answer ____
2. 221 × 59      Estimate _____        Exact Answer ____

## PART 8

*Some of the following problems contain errors. Find the errors by making an estimate to see if the answer is in the ballpark, and then using your calculator, correct the errors you find and write a sentence describing the errors.*

| 1. | 76 | 2. | 35 | 3. | 48 | 4. | 291 |
|---|---|---|---|---|---|---|---|
| | × 8 | | × 9 | | × 71 | | × 21 |
| | 568 | | 315 | | 48 | | 291 |
| | | | | | + 336 | | + 5820 |
| | | | | | 384 | | 6,111 |

Which problem(s) contain errors?

Write a description of each error and give the correct answers.

_____
_____
_____
_____

Assessments • Number Sense                Unit 3 Test

# ASSESSMENTS
## Quiz

Name _____  Date _____

## PART 1
*Solve the following basic and extended division facts.*

**1.** 14 ÷ 7 = ____        140 ÷ 7 = ____        1,400 ÷ 7 = ____

**2.** 28 ÷ 4 = ____        280 ÷ 4 = ____        2,800 ÷ 4 = ____

**3.** ____ ÷ 5 = 7         ____ ÷ 5 = 70         ____ ÷ 5 = 700

**4.** 36 ÷ ____ = 4        360 ÷ ____ = 40       3,600 ÷ ____ = 400

## PART 2
*Write multiplication/division fact families for the following.*

**1.** 6, 3, and 18          **2.** 9, 8, and 72          **3.** 7, 6, and 42

_____           _____           _____

_____           _____           _____

_____           _____           _____

_____           _____           _____

## PART 3
*Use your calculator to solve the following problems that have remainders. Round the answers to the nearest whole number.*

1. $5\overline{)26}$

   Calculator Answer: ____

   Rounded Answer: ____

2. $3\overline{)20}$

   Calculator Answer: ____

   Rounded Answer: ____

3. $7\overline{)34}$

   Calculator Answer: ____

   Rounded Answer: ____

## PART 4
*Estimate the answers to the following problems by finding the closest basic fact (the near fact). Then solve the near fact.*

1. $4\overline{)27}$

   Near Fact: ____)‾‾‾

2. $9\overline{)84}$

   Near Fact: ____)‾‾‾

3. $8\overline{)71}$

   Near Fact: ____)‾‾‾

4. $7\overline{)40}$

   Near Fact: ____)‾‾‾

# ASSESSMENTS
Test

Name _____  Date _____

## PART 1
Solve the following basic and extended division facts.

1. $56 \div 7 =$ ____     $560 \div 7 =$ ____     $5{,}600 \div 7 =$ ____
2. $36 \div 9 =$ ____     $360 \div 9 =$ ____     $3{,}600 \div 9 =$ ____
3. ____ $\div 6 = 7$       ____ $\div 6 = 70$       ____ $\div 6 = 70$
4. $12 \div$ ____ $= 4$    $120 \div$ ____ $= 40$   $1{,}200 \div$ ____ $= 400$

## PART 2
Write multiplication/division fact families for the following.

1. 6, 9, and 54     2. 4, 8, and 32     3. 7, 3, and 21

_____   _____   _____
_____   _____   _____
_____   _____   _____
_____   _____   _____

## PART 3
Use your calculator to solve the following problems that have remainders. Then round the answers to the nearest whole number.

1. $7\overline{)26}$     Calculator Answer: ____     Rounded Answer: ____
2. $9\overline{)20}$     Calculator Answer: ____     Rounded Answer: ____
3. $8\overline{)34}$     Calculator Answer: ____     Rounded Answer: ____

## PART 4
*Estimate the answers to the following problems by finding the closest basic fact (the near fact). Then solve the near fact.*

1. 9)86

   Near Fact:

2. 8)35

   Near Fact:

## PART 5
*Estimate the answers to the following problems by finding the closest fact (the near extended fact). Then solve the near extended fact.*

1. 7)279

   Near Extended Fact:

2. 6)433

   Near Extended Fact:

## PART 6
*Pull out the 10s in the following problems. Do not solve.*

1. 70)420

   Pull Out the 10s:

2. 60)360

   Pull Out the 10s:

## PART 7
*A student entered the following problem on a calculator.*

8)58

The student got the following answer:

.1379310344827S

1. Estimate what the answer should be by finding a near fact.

2. Enter the original problem on your calculator. What answer did you get?

3. What error did the student make?

_____
_____

# ASSESSMENTS
## Quiz 1

Name _____  Date _____

## PART 1
*Find the numbers shown by the following arrays. Then list all the factors for the number.*

1. The number is: _____

    Its factors are: _____

2. The number is: _____

    Its factors are: _____

3. The number is: _____

    Its factors are: _____

# PART 2
**List the factors for each of the following numbers.**

1. 12: _____
2. 15: _____
3. 23: _____
4. 25: _____

# PART 3
**Answer the following questions about prime numbers.**

1. What is a prime number? _____
_____

2. Which of the following numbers are prime? Circle them.

   1   2   3   5   7   9   11   15   21   23   29   30

3. Give an example of a prime number and draw its array(s).

# ASSESSMENTS
## Quiz 2

Name _____  Date _____

## PART 1

*For each of the following numbers, write which dividing rule(s) work for that number. Remember, these are the dividing rules:*

| 2 | We can divide a number by 2 if it is an even number. |
|---|---|
| 3 | We can divide a number by 3 if we can add up its digits and divide that number by 3 evenly. |
| 5 | We can divide a number by 5 if it ends in a 5 or a 0. |
| 6 | We can divide a number by 6 if we can divide it by 2 *and* by 3. |
| 10 | We can divide a number by 10 if it ends in a 0. |

| Number | Dividing Rule or Rules |
|---|---|
| 1. 12 | |
| 2. 16 | |
| 3. 20 | |
| 4. 30 | |

## PART 2
*Draw a prime factor tree for the following number.*

## PART 3
*Find the common factors of the following pairs of numbers.*

1. 8 | 1 | 2 | 3 | 4 | 5 | 6 | 7 | 8 |
   14 | 1 | 2 | 3 | 4 | 5 | 6 | 7 | 8 | 9 | 10 | 11 | 12 | 13 | 14 |

   The common factors are: _____

2. 16 | 1 | 2 | 3 | 4 | 5 | 6 | 7 | 8 | 12 | 16 | 20 | 24 |
   24 | 1 | 2 | 3 | 4 | 5 | 6 | 7 | 8 | 12 | 16 | 20 | 24 |

   The common factors are: _____

## PART 4
*Find the greatest common factor (GCF) of each of the following number pairs.*

1. 12 | 1 | 2 | 3 | 4 | 5 | 6 | 7 | 8 | 9 | 10 | 11 | 12 |
   14 | 1 | 2 | 3 | 4 | 5 | 6 | 7 | 8 | 9 | 10 | 11 | 12 | 13 | 14 |

   The GCF of 12 and 14 is: ____

2. 10 | 1 | 2 | 3 | 4 | 5 | 6 | 7 | 8 | 9 | 10 |
   13 | 1 | 2 | 3 | 4 | 5 | 6 | 7 | 8 | 9 | 10 | 11 | 12 | 13 |

   The GCF of 10 and 13 is: ____

# ASSESSMENTS
Test

Name _____  Date _____

## PART 1
*List all of the factors for the following numbers.*

1. 12 _____
2. 32 _____
3. 5 _____
4. 25 _____

## PART 2
*Answer the following questions about prime numbers.*

Which of the numbers in this list are prime? Circle them.

    2    5    6    7    9    11    13    15    23    27    33    37

What is a prime number? _____

_____

## PART 3
*Answer the following questions about square numbers.*

Which of the numbers in this list are square numbers? Circle them.

    1    2    4    8    9    11    13    16    23    25    32    36

What is a square number? _____

_____

# PART 4
*Find the area and the perimeter of the following figures. Assume each square on the grid is 1 cm by 1 cm.*

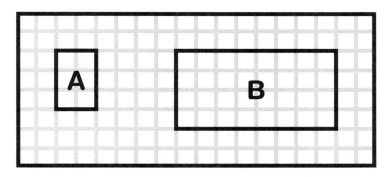

1. Area of A = _____
2. Perimeter of A = _____
3. Area of B = _____
4. Perimeter of B = _____

# PART 5
*Find the factors, common factors, and the GCF (greatest common factor) for each of these number pairs.*

1. 16 and 24

   Factors of 16: _____

   Factors of 24: _____

   Common Factors: _____

   GCF: _____

2. 9 and 18

   Factors of 9: _____

   Factors of 18: _____

   Common Factors: _____

   GCF: _____

## PART 6
*Find some multiples and the LCM (least common multiple) for each of these number pairs.*

1. 6 and 8

   Multiples of 6: _____

   Multiples of 8: _____

   LCM: _____

2. 3 and 4

   Multiples of 3: _____

   Multiples of 4: _____

   LCM: _____

## PART 7
*Draw factor trees for the following numbers. Be sure to circle the prime factors.*

          12                     15

          21

# PART 8
*Rewrite each of the following using exponents.*

1. 10 × 10 × 10 = _____
2. 2 × 2 × 2 × 2 = _____
3. 10 × 10 × 10 × 10 × 10 = _____

# PART 9
*Rewrite each of the following as a product of factors.*

1. $10^2$ = _____
2. $2^3$ = _____
3. $10^6$ = _____

# ASSESSMENT QUIZ AND TEST ANSWER KEY

## Unit 1 Quiz

### PART 1
1. 300
2. 5,000
3. 10
4. 0

### PART 2
1. 200 + 10 + 7
2. 300 + 70 + 5
3. 600 + 0 + 2
4. 400 + 50 + 0

### PART 3
1. 15   150   1,500
2. 6   60   600
3. 5   50   500

### PART 4

1. 
| | | |
|---|---|---|
| 200 | 20 | 5 |
| + 300 | 10 | 4 |
| 500 | 30 | 9 |

Answer: 539

2. 
| | | |
|---|---|---|
| 600 | 30 | 7 |
| + 100 | 50 | 1 |
| 700 | 80 | 8 |

Answer: 788

## Unit 1 Test

### PART 1
1. 10
2. 7
3. 400
4. 8,000

### PART 2
1. 300 + 40 + 7
2. 400 + 20 + 9
3. 500 + 0 + 8
4. 900 + 10 + 7

### PART 3
1. 11   110   1,100
2. 7   70   700
3. 8   80   800
4. 70   7   700
5. 70   700   7

### PART 4

1. 
| | | |
|---|---|---|
| 500 | 20 | 3 |
| + 100 | 40 | 3 |
| 600 | 60 | 6 |

Answer: 666

2. 
| | | |
|---|---|---|
| 700 | 20 | 1 |
| + 200 | 0 | 7 |
| 900 | 20 | 8 |

Answer: 928

3. 
| | | |
|---|---|---|
| 300 | 40 | 8 |
| + 400 | 10 | 7 |
| 700 | 50 | 15 |
| 700 | 60 | 5 |

Answer: 765

4. 
| | | |
|---|---|---|
| 400 | 0 | 9 |
| + 300 | 80 | 1 |
| 700 | 80 | 10 |
| 700 | 90 | 0 |

Answer: 790

### PART 5
The problem is asking how many students altogether attended the Homecoming dance and the Winter Festival dance.
Answer: 600

### PART 6
The answer to Problem 1 should be 291 (forgot the carry in the 10s place).

The answer to Problem 2 should be 365 (forgot the carry in the 10s place).

The answer to Problem 3 is correct.

The answer to Problem 4 should be 707 (forgot the carry in the 100s place).

# ASSESSMENT QUIZ AND TEST ANSWER KEY

## Unit 2 Quiz

### PART 1
1. $9 + 2 = 11$
   $2 + 9 = 11$
   $11 - 2 = 9$
   $11 - 9 = 2$
2. $7 + 6 = 13$
   $6 + 7 = 13$
   $13 - 6 = 7$
   $13 - 7 = 6$
3. $5 + 9 = 14$
   $9 + 5 = 14$
   $14 - 9 = 5$
   $14 - 5 = 9$

### PART 2
1. 
| 60 | 8 |
|---|---|
| − 30 | 4 |
| 30 | 4 |

   34

2. 
| 100 | 90 | 7 |
|---|---|---|
| − | 60 | 2 |
| 100 | 30 | 5 |

   135

3. 
| 500 | 70 | 9 |
|---|---|---|
| − 100 | 20 | 6 |
| 400 | 50 | 3 |

   453

### PART 3
1. Front-End
   Estimate: $800 - 200 = 600$
   Calculator Answer: 592
2. Quarter
   Estimate: $75 - 50 = 25$
   Calculator Answer: 19
3. Front-End
   Estimate: $800 - 100 = 700$
   Calculator Answer: 678
4. Quarter
   Estimate: $875 - 525 = 350$
   Calculator Answer: 348

### PART 4
1. 37
2. 691
3. 402
4. 109

## Unit 2 Test

### PART 1
1. 8  80  800
2. 8  80  800
3. 14  140  1,400
4. 6  60  600

### PART 2
1. $2 + 9 = 11$
   $9 + 2 = 11$
   $11 - 2 = 9$
   $11 - 9 = 2$
2. $20 + 90 = 110$
   $90 + 20 = 110$
   $110 - 90 = 20$
   $110 - 20 = 90$
3. $7 + 8 = 15$
   $8 + 7 = 15$
   $15 - 7 = 8$
   $15 - 8 = 7$
4. $70 + 80 = 150$
   $80 + 70 = 150$
   $150 - 70 = 80$
   $150 - 80 = 70$

### PART 3
1. 
| 90 | 7 |
|---|---|
| − 40 | 5 |
| 50 | 2 |

   52

2. 
| 800 | 70 | 9 |
|---|---|---|
| − | 50 | 8 |
| 800 | 20 | 1 |

   821

3. 
| 300 | 50 | 8 |
|---|---|---|
| − 100 | 20 | 7 |
| 200 | 30 | 1 |

   231

## PART 4
1. Quarter
   Estimate: 75 − 25 = 50
   Calculator Answer: 47
2. Front-End
   Estimate: 500 − 300 = 200
   Calculator Answer: 199
3. Quarter
   Estimate: 325 − 200 = 125
   Calculator Answer: 125
4. Front-End
   Estimate: 900 − 600 = 300
   Calculator Answer: 299

## PART 5
1. 23
2. 904
3. 360
4. 404

## PART 6
1. Yes, the problem has an error. The answer should be 19.
2. No, the problem does not have an error.
3. Yes, the problem has an error. The answer should be 192.
4. No, the problem does not have an error.

## PART 7
1. The best method is traditional subtraction because there is no regrouping. The answer is 222.
2. The best method is good number sense because the top number is only 2 more than the bottom number. The answer is 2.
3. The best method is estimation and calculator because there are several regroupings. The estimate is 1,000 − 600 = 400. The calculator answer is 438.

# ASSESSMENT QUIZ AND TEST ANSWER KEY

## UNIT 3

### Unit 3 Quiz
**PART 1**
1. 48     480     4,800
2. 63     630     6,300
3. 56     560     5,600
4. 45     450     4,500

**PART 2**
1. $4 \times 10$
2. $6 \times 10$
3. $50 \times 10$
4. $40 \times 10$
5. $60 \times 10$
6. $5 \times 10$
7. $400 \times 10$
8. $600 \times 10$
9. $500 \times 10$

**PART 3 (Items in bold type are already given in the table.)**

| The Starting Number | ? × 1,000 | ? × 100 | ? × 10 |
|---|---|---|---|
| 5,000 | 5 × 1,000 | **50 × 100** | **500 × 10** |
| 7,000 | 7 × 1,000 | **70 × 100** | 700 × 10 |
| 9,000 | 9 × 1,000 | 90 × 100 | **900 × 10** |
| 6,000 | 6 × 1,000 | **60 × 100** | 600 × 10 |

**PART 4**

1.      30
   ×    7
      63
   +  210
     273

2.      70
   ×    6
      48
   +  420
     468

3.      60
   ×    9
      45
   +  540
     585

4.      30
   ×    8
      72
   +  240
     312

### Unit 3 Test
**PART 1**
1. 42     420     4,200
2. 72     720     7,200
3. 56     560     5,600
4. 54     540     5,400

**PART 2**
1. $6 \times 10$
2. $2 \times 10$
3. $8 \times 10$
4. $9 \times 10$
5. $15 \times 10$
6. $30 \times 10$

**PART 3**
1. 20 mm or 2 cm
2. 50 cm or 5 dm
3. 100 dm or 10 m
4. 60 mm or 6 cm
5. 80 cm or 8 dm

**PART 4**

1.      90
   ×    6
      42
   +  540
     582

2.      80
   ×    9
      45
   +  720
     765

**PART 5 (Items in bold type are already given in the table.)**

| The Starting Number | ? × 1,000 | ? × 100 | ? × 10 |
|---|---|---|---|
| **1,000** | 1 × 1,000 | 10 × 100 | 100 × 10 |
| 2,000 | 2 × 1,000 | 20 × 100 | 200 × 10 |
| 3,000 | 3 × 1,000 | 30 × 100 | 300 × 10 |
| 6,000 | 6 × 1,000 | 60 × 100 | 600 × 10 |
| 7,000 | 7 × 1,000 | 70 × 100 | 700 × 10 |

**PART 6**
1. 288     2. 336     3. 648

**PART 7**
1. Estimate: $70 \times 8 = 560$
   Calculator Answer: 536
2. Estimate: $200 \times 60 = 12,000$
   Calculator Answer: 13,039

**PART 8**
The answer to Problem 1 should be 608 (forgot to add the carry).
Problem 2 is correct.
The answer to Problem 3 should be 3,408 (did not line up properly; forgot 0 place holder on 2nd line).
Problem 4 is correct.

# ASSESSMENT QUIZ AND TEST ANSWER KEY

## UNIT 4

## Unit 4 Quiz

### PART 1
1. 2     20     200
2. 7     70     700
3. 35     350     3,500
4. 9     9     9

### PART 2
1. $6 \times 3 = 18$
   $3 \times 6 = 18$
   $18 \div 6 = 3$
   $18 \div 3 = 6$
2. $9 \times 8 = 72$
   $8 \times 9 = 72$
   $72 \div 9 = 8$
   $72 \div 8 = 9$
3. $7 \times 6 = 42$
   $6 \times 7 = 42$
   $42 \div 6 = 7$
   $42 \div 7 = 6$

### PART 3
1. Calculator Answer: 5.2
   Rounded Answer: 5
2. Calculator Answer: 6.66666
   Rounded Answer: 7
3. Calculator Answer: 4.8571
   Rounded Answer: 5

### PART 4
1. The near fact for $27 \div 4 \rightarrow 28 \div 4 = 7$
2. The near fact for $84 \div 9 \rightarrow 81 \div 9 = 9$
3. The near fact for $71 \div 8 \rightarrow 72 \div 8 = 9$
4. The near fact for $40 \div 7 \rightarrow 42 \div 7 = 6$

## Unit 4 Test

### PART 1
1. 8     80     800
2. 4     40     400
3. 42     420     4,200
4. 3     3     3

### PART 2
1. $6 \times 9 = 54$
   $9 \times 6 = 54$
   $54 \div 6 = 9$
   $54 \div 9 = 6$
2. $4 \times 8 = 32$
   $8 \times 4 = 32$
   $32 \div 4 = 8$
   $32 \div 8 = 4$
3. $7 \times 3 = 21$
   $3 \times 7 = 21$
   $21 \div 7 = 3$
   $21 \div 3 = 7$

### PART 3
1. Calculator Answer: 3.714285714285
   Rounded Answer: 4
2. Calculator Answer: 2.2222222222222
   Rounded Answer: 2
3. Calculator Answer: 4.25
   Rounded Answer: 4

### PART 4
1. $9 \overline{)86}$    Near Fact: $9 \overline{)81}$ = 9
2. $8 \overline{)35}$    Near Fact: $8 \overline{)32}$ = 4

### PART 5
1. $7 \overline{)279}$    Near Extended Fact: $7 \overline{)280}$ = 40
2. $6 \overline{)433}$    Near Extended Fact: $6 \overline{)420}$ = 70

### PART 6
1. $70 \overline{)420}$
   Pull Out the 10s: $7 \times 10 \overline{)42 \times 10}$
2. $60 \overline{)360}$
   Pull Out the 10s: $6 \times 10 \overline{)36 \times 10}$

### PART 7
1. Estimate: $56 \div 8 = 7$
2. Calculator Answer: 7.25
3. The error was made by entering the numbers into the calculator in reverse order.
4. The student entered $8 \div 58$.

# ASSESSMENT QUIZZES AND TEST ANSWER KEY

## Unit 5 Quiz 1
### PART 1
1. The number is 12. Its factors are 1, 2, 3, 4, 6, and 12.
2. The number is 7. Its factors are 1 and 7.
3. The number is 9. Its factors are 1, 3, and 9.

### PART 2
1. 12: 1, 2, 3, 4, 6, and 12.
2. 15: 1, 3, 5, and 15.
3. 23: 1 and 23.
4. 25: 1, 5, and 25.

### PART 3
1. A prime number is a number with only two factors, 1 and itself.
2. The numbers from the list that are prime are 2, 3, 5, 7, 11, 23, and 29.

See individual student work. The number should be a prime number. It will have only one array.

## Unit 5 Quiz 2
### PART 1
1. Dividing rules 2, 3, and 6
2. Dividing rule 2
3. Dividing rules 2, 5, and 10
4. Dividing rules 2, 3, 5, 6, and 10

### PART 2
Students' trees may vary depending on the facts they used, but the prime factors will be the same.

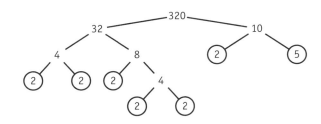

### PART 3
1. The common factors are 1 and 2.
2. The common factors are 1, 2, 4, and 8.

### PART 4
1. The GCF is 2.
2. The GCF is 1.

## Unit 5 Test
### PART 1
1. 12: 1, 2, 3, 4, 6, and 12
2. 32: 1, 2, 4, 8, 16, and 32
3. 5: 1 and 5
4. 25: 1, 5, and 25

### PART 2
The numbers from the list that are prime are 2, 5, 7, 11, 13, 23, 37.
A prime number is a number that has only two factors, 1 and itself.

### PART 3
The numbers from the list that are square numbers are 4, 9, 16, 25, and 36.
A square number is a number with a square array (equal number of rows and columns).

### PART 4
1. The area of A is $2 \times 3 = 6$ sq. cm.
2. The perimeter of A is $2 + 2 + 3 + 3 = 10$ cm.
3. The area of B is $8 \times 4 = 32$ sq. cm.
4. The perimeter of B is $8 + 8 + 4 + 4 = 24$ cm.

### PART 5
1. The factors of 16 are 1, 2, 4, 8, and 16.
   The factors of 24 are 1, 2, 3, 4, 6, 8, 12, and 24.
   The common factors are 1, 2, 4, and 8.
   The GCF is 8.
2. The factors of 9 are 1, 3, and 9.
   The factors of 18 are 1, 2, 3, 6, 9, and 18.
   The common factors are 1, 3, and 9.
   The GCF is 9.

## PART 6
1. Some multiples of 6 are 6, 12, 18, 24, 30, 36, 42, 48, 54, 60.
   Some multiples of 8 are 8, 16, 24, 32, 40, 48, 56, 64, 72, 80.
   The LCM is 24.
2. Some multiples of 3 are 3, 6, 9, 12, 15, 18, 21, 24, 27, 30.
   Some multiples of 4 are 4, 8, 12, 16, 20, 24, 28, 32, 36, 40.
   The LCM is 12.

## PART 7
Students' trees may vary depending on the facts they used, but the prime factors will be the same.

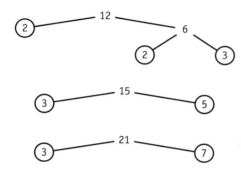

## PART 8
1. $10 \times 10 \times 10 = 10^3$
2. $2 \times 2 \times 2 \times 2 = 2^4$
3. $10 \times 10 \times 10 \times 10 \times 10 = 10^5$

## PART 9
1. $10^2 = 10 \times 10$
2. $2^3 = 2 \times 2 \times 2$
3. $10^6 = 10 \times 10 \times 10 \times 10 \times 10 \times 10$

# ASSESSMENT PRETESTS

# PRETEST ADDITION 1

Name _____  Date _____

| 1 | 4 | 0 | 8 | 7 |
|---|---|---|---|---|
| + 6 | + 4 | + 3 | + 2 | + 2 |

| 4 | 8 | 2 | 1 | 7 |
|---|---|---|---|---|
| + 2 | + 8 | + 5 | + 4 | + 1 |

| 2 | 8 | 2 | 6 | 3 |
|---|---|---|---|---|
| + 8 | + 1 | + 3 | + 1 | + 3 |

| 1 | 0 | 4 | 6 | 6 |
|---|---|---|---|---|
| + 3 | + 7 | + 1 | + 6 | + 2 |

| 9 | 5 | 2 | 1 | 8 |
|---|---|---|---|---|
| + 1 | + 5 | + 1 | + 8 | + 0 |

| 1 | 1 | 2 | 9 | 5 |
|---|---|---|---|---|
| + 2 | + 5 | + 2 | + 9 | + 2 |

| 7 | 2 | 1 | 6 | 3 |
|---|---|---|---|---|
| + 7 | + 7 | + 1 | + 0 | + 1 |

| 5 | 3 | 2 | 7 | 2 |
|---|---|---|---|---|
| + 1 | + 2 | + 6 | + 1 | + 4 |

Assessments • *Number Sense*   Pretest   **41**

# PRETEST ADDITION 2

Name _____  Date _____

| | | | | |
|---|---|---|---|---|
| 3<br>+ 7 | 4<br>+ 8 | 7<br>+ 9 | 6<br>+ 5 | 5<br>+ 8 |
| 9<br>+ 6 | 4<br>+ 7 | 3<br>+ 9 | 5<br>+ 7 | 6<br>+ 3 |
| 8<br>+ 5 | 3<br>+ 5 | 7<br>+ 3 | 9<br>+ 2 | 8<br>+ 7 |
| 6<br>+ 4 | 8<br>+ 3 | 9<br>+ 5 | 6<br>+ 7 | 7<br>+ 5 |
| 4<br>+ 3 | 6<br>+ 8 | 5<br>+ 9 | 4<br>+ 6 | 8<br>+ 4 |
| 2<br>+ 9 | 5<br>+ 3 | 8<br>+ 9 | 3<br>+ 6 | 7<br>+ 4 |
| 9<br>+ 4 | 3<br>+ 8 | 9<br>+ 3 | 7<br>+ 8 | 6<br>+ 9 |
| 9<br>+ 8 | 7<br>+ 6 | 9<br>+ 7 | 3<br>+ 4 | 8<br>+ 6 |

# PRETEST SUBTRACTION 1

Name _____    Date _____

| 8   | 4   | 3   | 7   | 6   |
|-----|-----|-----|-----|-----|
| −0  | −2  | −3  | −5  | −1  |

| 5   | 10  | 9   | 8   | 5   |
|-----|-----|-----|-----|-----|
| −3  | −8  | −7  | −8  | −2  |

| 3   | 8   | 5   | 10  | 6   |
|-----|-----|-----|-----|-----|
| −2  | −2  | −0  | −2  | −3  |

| 4   | 8   | 9   | 4   | 8   |
|-----|-----|-----|-----|-----|
| −1  | −4  | −9  | −3  | −6  |

| 8   | 6   | 9   | 5   | 7   |
|-----|-----|-----|-----|-----|
| −7  | −2  | −1  | −5  | −1  |

| 10  | 9   | 3   | 2   | 8   |
|-----|-----|-----|-----|-----|
| −10 | −2  | −1  | −2  | −1  |

| 2   | 7   | 6   | 7   | 10  |
|-----|-----|-----|-----|-----|
| −1  | −6  | −5  | −2  | −9  |

| 6   | 6   | 10  | 5   | 9   |
|-----|-----|-----|-----|-----|
| −6  | −4  | −1  | −4  | −8  |

Assessments • Number Sense

# PRETEST SUBTRACTION 2

Name _____  Date _____

| 13 | 11 | 7 | 9 | 13 |
|---|---|---|---|---|
| − 8 | − 9 | − 4 | − 5 | − 7 |

| 13 | 7 | 11 | 8 | 12 |
|---|---|---|---|---|
| − 5 | − 3 | − 6 | − 3 | − 6 |

| 9 | 13 | 10 | 14 | 9 |
|---|---|---|---|---|
| − 4 | − 6 | − 4 | − 8 | − 6 |

| 14 | 16 | 13 | 15 | 10 |
|---|---|---|---|---|
| − 5 | − 9 | − 9 | − 6 | − 7 |

| 12 | 15 | 14 | 12 | 15 |
|---|---|---|---|---|
| − 7 | − 8 | − 6 | − 8 | − 9 |

| 11 | 12 | 13 | 16 | 8 |
|---|---|---|---|---|
| − 8 | − 3 | − 4 | − 8 | − 5 |

| 17 | 10 | 12 | 15 | 14 |
|---|---|---|---|---|
| − 9 | − 3 | − 9 | − 7 | − 9 |

| 11 | 17 | 12 | 10 | 16 |
|---|---|---|---|---|
| − 7 | − 8 | − 5 | − 6 | − 7 |

# PRETEST MULTIPLICATION 1

Name _____     Date _____

| 9 × 1 | 10 × 3 | 7 × 1 | 0 × 4 | 10 × 7 |
|---|---|---|---|---|
| 5 × 7 | 9 × 5 | 6 × 1 | 3 × 5 | 6 × 10 |
| 10 × 6 | 1 × 4 | 5 × 8 | 2 × 2 | 4 × 1 |
| 5 × 9 | 10 × 4 | 10 × 0 | 8 × 5 | 10 × 2 |
| 3 × 1 | 10 × 10 | 7 × 5 | 5 × 3 | 2 × 5 |
| 10 × 8 | 2 × 3 | 5 × 5 | 1 × 8 | 10 × 9 |
| 2 × 4 | 9 × 2 | 2 × 10 | 1 × 6 | 2 × 7 |
| 5 × 4 | 7 × 10 | 6 × 5 | 10 × 5 | 8 × 2 |

# PRETEST MULTIPLICATION 2

Name _____  Date _____

| 4 × 9 | 6 × 7 | 3 × 7 | 6 × 9 | 7 × 8 |
|---|---|---|---|---|
| 7 × 4 | 3 × 9 | 5 × 5 | 8 × 3 | 4 × 8 |
| 8 × 8 | 6 × 4 | 8 × 7 | 7 × 3 | 6 × 7 |
| 3 × 3 | 6 × 8 | 6 × 3 | 7 × 9 | 9 × 3 |
| 3 × 4 | 9 × 6 | 8 × 6 | 3 × 8 | 4 × 4 |
| 7 × 6 | 4 × 6 | 4 × 7 | 9 × 7 | 8 × 4 |
| 4 × 9 | 8 × 9 | 6 × 6 | 4 × 3 | 8 × 8 |
| 5 × 4 | 9 × 9 | 3 × 6 | 7 × 7 | 9 × 8 |